NISTIR 7461

Heat and Mass Transfer Through Gypsum Partitions Subjected to Fire Exposures

Scott Kukuck
Building and Fire Research Laboratory

August 2009

U.S. Department of Commerce
Gary Locke, Secretary

National Institute of Standards and Technology
Patrick D. Gallagher, Deputy Director

ABSTRACT

A model is presented that describes the heat transfer through a gypsum wallboard partition assembly incorporating the mass transport effects of water in liquid and vapor form. Sources of water include surface bound (adsorbed) water and hydrated water that is chemically bound with the crystal matrix of gypsum. Liberated water is allowed to migrate through the porous structure through molecular diffusion and pressure driven flow. Evaporation or condensation occurs when the partial pressure of water vapor with the pore space is less than or greater than the saturation pressure, respectively. Results obtained from the model are compared to measurements taken during a standard fire resistance test. It is found that the surface temperatures are typically under-predicted, although qualitatively similar behavior is observed. The analysis implies that the liberation and transport of water in both its liquid and vapor form plays a significant role in the thermal response of gypsum wallboard subjected to fire exposures, perhaps even beyond the extent to which the current model has been developed. The model can be used, under the assumptions through which it was developed, to conduct sensitivity studies of the physical parameters and evaluate the effect upon the fire resistance of the system.

ACKNOWLEDGEMENTS

The author wishes to acknowledge Samuel Manzello, Richard Gann, Kuldeep Prasad and Walter Jones of the Fire Research Division for their assistance during the formulation of the model and discussions about the physical response of gypsum board subjected to fire exposures.

DISCLAIMER

Certain companies and commercial products are identified in this paper in order to specify adequately the source of information or of equipment used. Such identification does not imply endorsement or recommendation by the National Institute of Standards and Technology, nor does it imply that this source or equipment is the best available for the purpose.

POLICY OF NIST REGARDING THE INTERNATIONAL SYSTEM OF UNITS

The policy of NIST is to use the International System of Units (metric units) in all publications. In this document, however, units are presented in metric units or the inch-pound system, whichever is prevalent to the discipline.

TABLE OF CONTENTS

	page
ABSTRACT	iii
ACKNOWLEDGEMENTS	iv
DISCLAIMER	iv
POLICY OF NIST REGARDING THE INTERNATIONAL SYSTEM OF UNITS	iv
INTRODUCTION	1
MODEL DEVELOPMENT	4
GYPSUM WALLBOARD	4
GAS LAYERS	7
BOUNDARY CONDITIONS	9
SOLUTION TECHNIQUE	10
MODEL PARAMETERS	11
DENSITY	11
SPECIFIC HEAT	11
THERMAL CONDUCTIVITY	11
POROSITY	12
DIFFUSIVITY	12
PERMEABILITY	13
ADSORPTION	13
REACTIONS	13
GAS PROPERTIES	14
WATER PROPERTIES	14
RESULTS	14
CONCLUSIONS & FUTURE WORK	18
REFERENCES	21

Heat and Mass Transfer Through Gypsum Partitions Subjected to Fire Exposures

INTRODUCTION

Traditional fire resistance testing in the United States has been based upon ASTM standard E119, "Standard Test Methods for Fire Tests of Building Construction and Materials" [1]. In this test, building components are subjected to a furnace exposure intended to represent a standard fire. The components are then rated, with units of time, on their ability to withstand the exposure up to some defined criteria. The results of the test have been typically placed in prescriptive building codes that mandate specific ratings for construction assemblies. Issues with this approach revolve around the inability to predict the response of the construction to fires that almost certainly will not match the standard fire simulated in the furnace. Thus while construction A may achieve a rating twice as long as construction B in the furnace exposure, it is not possible to predict how long these constructions will survive, either absolutely or relatively, under a real exposure from simply the test measurements. Furthermore, prescribed ratings for all components of a building design may result in excessive protection of portions of the construction that may never see exposures matching the standard fire or perhaps, more dangerously, insufficient protection of critical components that may see exposures exceeding that of the standard fire.

With these issues in mind, some authorizing bodies have begun to implement the use of performance-based criteria in their building codes. Under this approach, designs are assessed on how they would perform during actual exposures. The caveat to this approach is that it is not feasible, neither practically nor economically, to test in the full scale all possible exposures a building design may be subjected to. What is needed for this approach are models capable of accurately predicting the response of construction assemblies to a wide range of various conditions. These models could draw upon a small subset of full and reduced scale tests to yield the predicted response.

In the following work we focus on modeling one of the most common means of reducing fire growth and spread, i.e. compartmentation. Specifically, we deal with the response of gypsum partitions subjected to fire exposures. While models do currently exist [2-4], most utilize a single energy conservation equation. To approximate the effects of mass transport through and chemical dehydration of the materials that comprise the core of the gypsum wallboard, the thermal properties of the core material are typically modified in some manner. These models have shown success in predicting results from standard exposures. It is unclear, however, how the property modifications of these models respond to non-standard exposures. Mixed success has been observed in cases where comparisons of model results have been made with measurements from nonstandard fires.

The core material of gypsum wallboard is a porous solid composed primarily of calcium sulfate dihydrate ($CaSO_4.2H_2O$), a naturally occurring mineral in which two water molecules are chemically bound for every one calcium sulfate molecule within the crystal matrix. The presence of the water molecules is a key feature in establishing the fire resistance properties of gypsum. When heated, crystalline gypsum dehydrates and water is liberated, typically in two separate reversible chemical reactions [5],

$$CaSO_4.2H_2O + Q \rightarrow CaSO_4.\frac{1}{2}H_2O + \frac{3}{2}H_2O \qquad (1)$$

$$CaSO_4.\frac{1}{2}H_2O + Q \rightarrow CaSO_4 + \frac{1}{2}H_2O \qquad (2)$$

Both of these dehydration reactions are endothermic, i.e. they absorb energy, and therefore retard the flow of heat and enhance the fire resistance. These reactions generally occur at temperatures of between 125 °C and 225 °C.

A third reaction occurs at even higher temperature when the molecular structure of the soluble crystal reorganizes itself into a lower insoluble energy state,

$$CaSO_4(sol) \rightarrow CaSO_4(insol) + Q \qquad (3)$$

In contrast to the dehydration, this reaction is exothermic, i.e. releases energy. While the amount of energy released is not a significantly large amount, the molecular restructuring may eventually play an important role in establishing conditions for cracking of the gypsum board. This reaction occurs at temperature of around 400 °C, a temperature which also corresponds to a significant contraction of gypsum boards [6].

To demonstrate the importance of these reactions, the results of a Differential Scanning Calorimeter (DSC) scan is presented in figure 1 for the material comprising the core of a Type-X gypsum board [7]. In the figure, the specific heat with respect to the initial specimen mass is plotted as a function of temperature. Clearly visible on the graph are the two peaks from the dehydration reactions as well as the molecular restructuring reaction. The large contribution to the apparent specific heat from the dehydrations demonstrates the need to incorporate reaction effects into any heat transfer model for gypsum wallboard.

Including simply the thermal effects does not complete the story, however. A simple mass balance shows that $CaSO_4.2H_2O$ is composed of approximately 21 % by mass water. A gypsum core containing 85 % by mass $CaSO_4.2H_2O$ and having a density of 650 kg / m^3 therefore has a potential water density of 116 kg/m^3. It is easy to show that if this water remains completely *in situ* that it would necessarily exist in the saturated state. This would result in the internal pore pressure of the board being at the saturation pressure. At 250 °C this is 4.0 MPa (39 atm. Or 580 psi) and by 350 °C the pore pressures would be 16.5 MPa (163 atm. Or 2400 psi). The porous nature of gypsum board necessitates that these pressures result in at least some migration of the water vapor. This migration is not trivial, however, and may present its own unique challenge. If liberated water vapor migrates to a cooler region it may condense, releasing energy and increasing the local temperature. As heat continues to be transported across the material this water will eventually evaporate again, continuing the cycle through the width of the board. While in its liquid state, the relatively large thermal conductivity of water will increase the flux of heat conducted through the region.

An additional source of water that exists for gypsum board is due to adsorption effects. Adsorption is a weak bonding of liquid water to the pore surfaces. The amount of water adsorbed is a weak function of temperature, and a strong function of the relative humidity [8]. Although

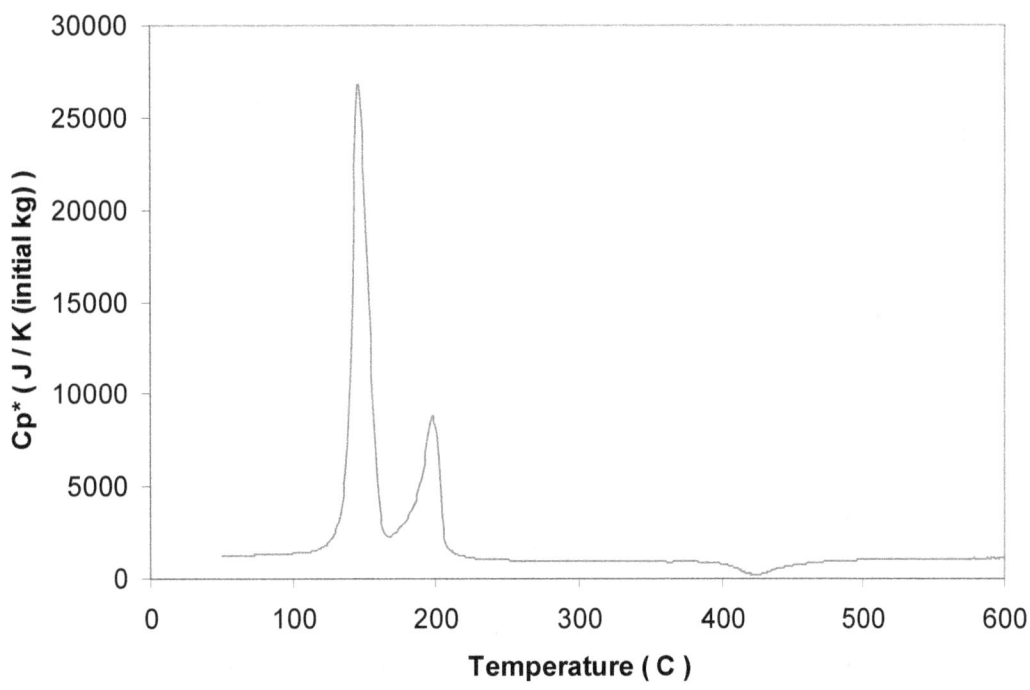

Figure 1: Apparent specific heat of 0.0159 m thick Type-X gypsum wallboard [7].

the overall amount of adsorbed water is typically small, approximately 3 %, energy is required to liberate it, further retarding the flow of heat. As in case of the hydrated water, once liberated the de-adsorbed water vapor will also migrate and potentially condense further along during the exposure.

We present a model that incorporates not only the heat transfer across partitions, but also includes the effects of mass transport through the materials comprising the partition. State and constitutive relations are included that simulate the dehydration, adsorption and condensation / evaporation of water. Bulk mass transport arises from pressure driven flow through a porous substance utilizing Darcy's Law. Molecular diffusion is also included in the model. Results obtained using the model are checked against test data from a standard fire resistance test [4,9]. Still in the development stage, current model results show qualitatively similar behavior to test measurements, however the model under predicts the temperature response of the partition. Analysis implies that the transport of moisture does in fact play a significant role in the thermal response of gypsum partitions subjected to fire exposure and that transport may well go beyond simply pressure driven flow. Although the prediction of temperatures is not yet accurate, useful information can still be obtained from the model. These results will be discussed as will future avenues of research intended to improve and expand the model.

Model Development

The basic construction being considered in this work is displayed schematically in figure 2. Here, layers of gypsum wallboard are attached on either side of a series of wall studs currently considered to be of arbitrary composition. Although the model will be derived in general for any given number of dimension, in this initial phase of model development we have restricted attention to the thermal response of the gypsum wallboard and therefore will consider a one-dimensional model as outlined in the figure. The following sections outline the derivation of the model equations for: a) the gypsum wallboard, b) the gas layer between the wallboards arising from the studs and c) the boundary conditions. Following the derivations, a brief section describing the techniques utilizing in solving the series of derived equations is presented.

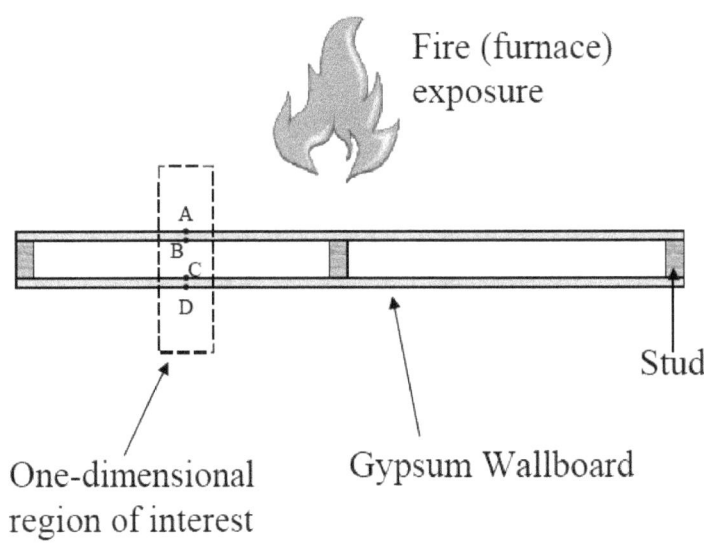

Figure 2: A schematic representation of the gypsum partition considered in the model

(1) Gypsum Wallboard

The governing equations utilized to model the transport of heat and mass across gypsum boards are obtained from conservation equations adapted for use in a porous substance. Darcy's law for bulk average flow through a porous material therefore supplants the momentum equation. As we are interested in heat and mass transfer, it is necessary to express conservation equations for each of the possible gas species as well as for any liquid phase water that may arise.

The gas conservation equations are expressed in molar form as,

$$\frac{\partial}{\partial t}\{\varphi_0 f_g C_1\} + \nabla \cdot \{f_g C_1 \mathbf{V}_g\} = \nabla \cdot \left(f_g C D \nabla \frac{C_1}{C} \right) \tag{4}$$

$$\frac{\partial}{\partial t}\{\varphi_0 f_g C_2\} + \nabla \cdot \{f_g C_2 \mathbf{V}_g\} = \nabla \cdot \left(f_g CD \nabla \frac{C_2}{C}\right) - \frac{1}{M_2}\frac{\partial}{\partial t}\{S_W + W + A\} \quad (5)$$

where in the above expressions: f_g is the pore fraction available for the gas, C_1 is the molar density of air in the pore space, C_2 is the molar density of water vapor in the pore space, C is the total gas molar density in the pore space and $\mathbf{V}_g$ is the bulk average velocity of the gas. Source terms in these equations arise from the liberation of hydrated water S_W, the evaporation / condensation of free water W, and the adsorption / desorption of surface bound water A. Parameters and constants in these expressions are the unsaturated porosity of the material φ_0, the unsaturated diffusivity of water vapor D and the molecular weight of water M_2.

In a similar fashion we can write the conservation of liquid water in mass form as

$$\frac{\partial}{\partial t}\{\varphi_0 f_\ell \rho_\ell\} + \nabla \cdot \{f_\ell \rho_\ell \mathbf{V}_\ell\} = \frac{\partial W}{\partial t} \quad (6)$$

where f_ℓ is the pore fraction containing liquid water, ρ_ℓ is the density of water and $\mathbf{V}_\ell$ is the bulk average velocity of the liquid.

As stated above, the bulk average velocities for both the liquid and gases are obtained utilizing Darcy's law,

$$\mathbf{V} = -\frac{K}{\mu}\nabla p \quad (7)$$

where K is the permeability, μ is the viscosity and p is the pressure.

The final conservation equation governs the transport of energy across the system. If we neglect energy transport in the gas phase due to convective effects and assume that the internal energy and enthalpy of the solid and gas phases are approximately equal, then we obtain,

$$S\frac{\partial h_S}{\partial t} + W\frac{\partial h_\ell}{\partial t} + A\frac{\partial h_A}{\partial t} + f_\ell \rho_\ell \mathbf{V}_\ell \cdot \nabla h_\ell$$
$$= \nabla \cdot \{k_c \nabla T\} - \left(\Delta h_S \frac{\partial S}{\partial t} + \Delta h_W \frac{\partial W}{\partial t} + \Delta h_A \frac{\partial A}{\partial t}\right) \quad (8)$$

where S is the density of the underlying porous solid, h_S is the enthalpy of the porous solid, h_ℓ is the enthalpy of the free liquid water and h_A is the enthalpy of the adsorbed water. The terms Δh_S, Δh_W and Δh_A correspond to the energy required for the changes in state / phase associated with the dehydration of water from the gypsum, the condensation of free liquid water and the adsorption of surface bound water, respectively. The remaining parameter on the right hand side of the equation is the composite thermal conductivity k_c, which includes not only the conductivity of the solid material but also includes a contribution from the presence of any liquid phase water and is approximated by the expression,

$$k_c = k_S + \varphi_0 f_\ell k_\ell \quad (9)$$

with k_s the thermal conductivity of the solid and k_ℓ the thermal conductivity of liquid water. We now introduce the concept of specific heat to obtain the final form of the energy equation,

$$\{SC_S + WC_\ell + AC_A\}\frac{\partial T}{\partial t} + f_\ell \rho_\ell C_\ell \mathbf{V}_\ell \cdot \nabla T$$
$$= \nabla \cdot \{k_c \nabla T\} - \left(\Delta h_S \frac{\partial S}{\partial t} + \Delta h_W \frac{\partial W}{\partial t} + \Delta h_A \frac{\partial A}{\partial t}\right) \quad (10)$$

with C_S, C_ℓ and C_A the specific heats of the porous solid, free liquid water and adsorbed water respectively and T the temperature.

The conservation equations are supplemented with appropriate equations to fully define the state of the gas and liquid. Each gas phase component is considered to be an ideal gas,

$$p_1 = C_1 R_0 T \quad (11)$$
$$p_2 = C_2 R_0 T \quad (12)$$
$$p = p_1 + p_2 \quad (13)$$

where p_i is the partial pressure of the ith gas component, p is the total pressure and R_0 is the universal gas constant. Condensation / evaporation is governed by partial pressure of the water vapor, p_2, and the saturation pressure which is a function of temperature only,

$$p_{sat} = p_{sat}(T). \quad (14)$$

With the conservation and state equations so defined, we turn now to the evaluation of specific terms. For the absorption term, we recognize that although the amount of adsorbed moisture depends upon the temperature, this dependence is typically weak. We chose therefore to neglect the temperature dependence and allow the amount of absorbed moisture to depend upon the local relative humidity only. The main purpose of including the effect of adsorption is to incorporate the initial presence of the surface bound water. We therefore will also introduce the approximation that adsorption is only important for the core material in its initial, and not dehydrated state. The equation that will govern the amount of adsorbed moisture then is,

$$A = A(\phi) \quad (15)$$

where ϕ is the local relative humidity. This isotherm can be measured and then a functional correlation introduced that allows us to define the amount of adsorbed moisture present for a pore space at a given local relative humidity.

For the chemical reaction terms, we recognize from figure 1 that there may exist for the core material the possibility of four unique states, each with its own thermal properties. We chose therefore to define a constitutive variable for each state, Γ_i, that ranges from zero to one. If Γ_i is identically zero, then there does not exist any solid in state i at that location. When Γ_i is equal to one, then the entire solid is composed of only the ith state. Transitions from one state to the next are considered to be reactions of an Arrhenius form that we write as,

$$\frac{\partial \Gamma}{\partial t} = A\Gamma^n e^{-B/T} \quad (16)$$

where A is a pre-exponential term (necessarily negative for the above form), B is similar to the activation energy of gas phase kinetics, and n is the reaction order. The dehydration reactions for gypsum are complex, and in actuality depend upon not only the amount of the initial state present (Γ) but also on the partial pressure of water vapor surrounding the substance (p_2) [5]. The level of detail required to completely define the reaction including this fact does not exist in the literature, however, and so we chose the above form as the model for the reactions. The presence of any material in a lower state (e.g. state 2) will depend first upon there being a reaction of the next higher state (e.g. state 1). The complete system of equations for the possible states of the core material is expressed therefore as,

$$\frac{\partial \Gamma_1}{\partial t} = A_1 \Gamma_1^{n_1} e^{-B_1/T} \tag{17}$$

$$\frac{\partial \Gamma_2}{\partial t} = A_2 \Gamma_2^{n_2} e^{-B_2/T} - A_1 \Gamma_1^{n_1} e^{-B_1/T} \tag{18}$$

$$\frac{\partial \Gamma_3}{\partial t} = A_3 \Gamma_3^{n_3} e^{-B_3/T} - A_2 \Gamma_2^{n_2} e^{-B_2/T} \tag{19}$$

$$\frac{\partial \Gamma_4}{\partial t} = \qquad - A_3 \Gamma_3^{n_3} e^{-B_3/T}. \tag{20}$$

The introduction of the constitutive variables for each state, Γ_i, further allows us to define the thermal properties of the solid material as a sum of the properties for each of the individual states,

$$S = \Gamma_1 \rho_{S1} + \Gamma_2 \rho_{S2} + \Gamma_3 \rho_{S3} + \Gamma_4 \rho_{S4} \tag{21}$$

$$SC_S = \Gamma_1 \rho_{S1} C_{S1} + \Gamma_2 \rho_{S2} C_{S2} + \Gamma_3 \rho_{S3} C_{S3} + \Gamma_4 \rho_{S4} C_{S4} \tag{22}$$

$$k_S = \Gamma_1 k_{S1} + \Gamma_2 k_{S2} + \Gamma_3 k_{S3} + \Gamma_4 k_{S4} \tag{23}$$

where ρ_{Si} is the density, C_{Si} is the specific heat and k_{Si} is the thermal conductivity of the ith state.

(2) Gas Layers

We treat the gas layers created by the studs as a lumped control volume at a uniform state, and therefore assume a well-mixed gas volume. We also assume that although the wallboards are securely fastened to the underlying structural framework that there exists sufficient leak points so as to maintain a constant pressure with the gas layer. A simplified schematic for this volume is presented in figure 3.

Conservation of species applied to the control volume yields the following equations,

$$V \frac{\partial C_1}{\partial t} = \oiint_A C_1'' dA - V \dot{C}_{1e} \tag{24}$$

$$V \frac{\partial C_2}{\partial t} = \oiint_A C_2'' dA - V \dot{C}_{2e} \tag{25}$$

and conservation of energy the following

$$V \frac{\partial}{\partial t} \{C_1 u_1 + C_2 u_2\} = \oiint_A (C_1'' h_1 + C_2'' h_2 + q'') dA - V \{\dot{C}_{1e} h_{1e} + \dot{C}_{2e} h_{2e}\}. \tag{26}$$

In the above equations V is the volume of the gas layer, A the associated area of the boundaries of the gas layer, C_1 the molar density of air molecules, C_2 the molar density of water vapor molecules and u_i the molar specific internal energy of the ith gas species. The flux terms C_1'' and C_2'' correspond to the mass flux that enters the control volume through the wallboard boundaries with corresponding molar specific enthalpy h_i. The terms $\dot{C}_{1e}$ and $\dot{C}_{2e}$ as h_{1e} and h_{2e} are the molar densities and molar specific enthalpies associated with the flow of gas that must exit (or enter) the control volume so as to satisfy the constant pressure constraint. We assume that the gas

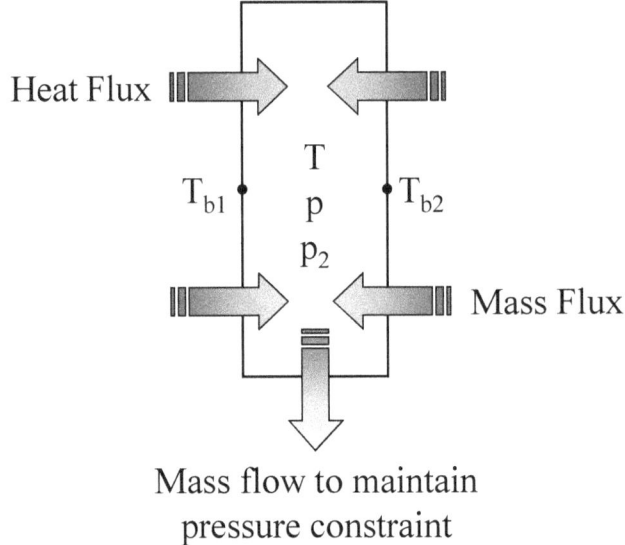

Figure 3: A simplified schematic of the control volume representation for the gas layer

comprising the control volume is transparent to radiation and therefore the heat flux q'' is simply due to convection effects from the wallboard boundaries to the gas.

The mass flux terms included in the above equations may arise from multiple sources. The first and most obvious is the flow of gas from the porous solid to the gas layer. The second source that may arise is from the evaporation / condensation of liquid water from the boundary of the wallboard. We assume that this layer (if it exists) is small and does not intrude upon the control volume. Furthermore, any heat gain or loss associated with the liquid change of state is applied to the solid and not to the gas. Lastly, if the associated flux is positive, i.e. mass is entering the control volume, then the enthalpy h_i is evaluated at the boundary temperature. If the flux is negative, i.e. mass is leaving the control volume, then the enthalpy h_i is evaluated at the control volume temperature.

These equations are again supplemented by the appropriate state equations for a multi-component ideal gas,

$$p_1 = C_1 R_0 T \tag{27}$$
$$p_2 = C_2 R_0 T \tag{28}$$
$$p = p_1 + p_2. \tag{29}$$

We will also introduce specific heats for the gases to express both the internal energy and the enthalpy through,

$$h - h_0 = C_p(T - T_0) \tag{30}$$
$$u - u_0 = C_v(T - T_0) \tag{31}$$

where C_p is the specific heat of the gas at constant pressure and C_v is the specific heat of the gas at constant volume. The subscript 0 in the above expressions denote a suitable reference state.

(3) Boundary Conditions

We now discuss appropriate boundary conditions for use in the model. Because the model incorporates not only heat but also mass transfer, the number of applied conditions increases as it is necessary to supply information regarding the state and composition of the gases on either side of the porous wallboard. We distinguish between external (absolute) boundary conditions representing the fire exposure and ambient, and internal boundary conditions at gas layer / wallboard interfaces.

External Boundaries

As in the air layers we assume that the gas surrounding the wallboard is well mixed. Thus for the mass conservation conditions on the external boundaries, we apply pressure (partial and total) conditions at the interface with the values equal to those of the gas. We further assume that the total pressure in both the fire and in the ambient is maintained at a constant value,

$$p|_{boundary} = p_0 \tag{30}$$

$$p_2|_{boundary} = p_2(t) \tag{31}$$

For the thermal boundary conditions, heat flux is applied to the wallboard through both convective and radiative mechanisms. For the convective condition we introduce an expression appropriate for natural convection on a vertical surface,

$$q''_{convective} = h(T_0 - T_b) \tag{32}$$

with the heat transfer coefficient evaluated according to the appropriate correlation, c.f. [10]. In the above expression, h is the effective heat transfer coefficient, T_0 is the gas temperature and T_b is the boundary surface temperature. We recognize that while the well-mixed assumption of the gas layer implies a sufficient velocity distribution to stir the gas, this may negate the assumption of natural convection. Unfortunately, there is typically insufficient information provided to evaluate a forced convection correlation. Furthermore, while there are convective influences, especially early in the exposure and at the ambient boundary, the dominant mechanism for heat transfer throughout the majority of the exposure will be due primarily to radiation and will quickly overshadow the convection.

For the radiative flux contribution, we note that there may be both an applied flux independent of the gas condition, e.g. for radiant panel exposures during testing or perhaps walls on the opposite side of the compartment that are at a different temperature, as well as contribution from the surrounding gas in fire exposures. We choose therefore to include separate terms for each of these possible radiant sources,

$$q''_{radiative} = q''_{ext}(t) + \varepsilon_{eff} \sigma (T_0^4 - T_b^4) \tag{33}$$

where ε_{eff} is the effective emissivity that incorporates the emissivities of the surrounding gas and that of the porous solid.

By applying an infinitesimally thin control volume at the boundary interface, the thermal boundary condition is derived from the conservation of energy at the boundary and can be expressed as,

$$-k_{comp}\nabla T \cdot \hat{n}\big|_{boundary} = \varepsilon_{eff}\sigma(T_0^4 - T_b^4) + h(T_0 - T_b) + q''_{ext}. \quad (34)$$

Internal Boundaries

The application of boundary conditions for the internal gas layers is similar to that presented above for the external boundaries. We assume, however, that the gas layer is transparent to radiation, and therefore the heat flux at the two opposing surfaces is dependent upon the temperature of each surface. In a similar manner to that presented above for the external boundary conditions, the equations expressing conditions at the internal boundaries are given to be,

$$p\big|_{boundary\ 1} = p\big|_{boundary\ 2} = p_g \quad (35)$$

$$p_2\big|_{boundary\ 1} = p_2\big|_{boundary\ 2} = p_{2g} \quad (36)$$

$$-k_{comp}\nabla T \cdot \hat{n}\big|_{boundary\ 1} = \varepsilon_{eff}\sigma(T_{b1}^4 - T_{b2}^4) + h_1(T_{b1} - T_g) \quad (37)$$

$$-k_{comp}\nabla T \cdot \hat{n}\big|_{boundary\ 2} = \varepsilon_{eff}\sigma(T_{b1}^4 - T_{b2}^4) + h_2(T_g - T_{b2}) \quad (38)$$

where the terms p_g, p_{2g} and T_g represent the pressure, partial pressure of water vapor and temperature of the gas layer, respectively. The effective emissivity in the above expressions now depends upon the emissivities of each of the two surfaces, and is evaluated from the expression,

$$\varepsilon_{eff} = \left(\frac{1}{\varepsilon_1} + \frac{1}{\varepsilon_2} - 1\right)^{-1}. \quad (39)$$

The convective heat transfer coefficients, h_1 and h_2, are evaluated utilizing the same natural convection correlations as for the external boundaries.

(4) Solution Technique

The complex and nonlinear nature of the presented equations for modeling the heat and mass transfer through gypsum partitions necessitates the use of numerical methods in obtaining solutions. We implement a marker and cell (MAC) approach with the solid material divided into a finite number of cells. Vector quantities are evaluated at cell boundaries while scalar terms are evaluated at cell centers. Diffusive terms are calculated using a second-order central difference scheme while convective terms utilize an upwind biased approach.

To reduce as much as possible issues of stability, we adopt a backward Euler approach for the temporal derivatives. While this method is unconditionally stable for linear equations, the nonlinear nature of the equations as well as the discontinuities arising from liquid / gas phase changes require the use of a stability criteria similar to the CFL condition observed for explicit schemes,

$$\Delta t \leq \frac{\Delta x}{|\mathbf{V}_{max}|} \quad (40)$$

where $\mathbf{V}_{max}$ is the maximum convective velocity of the gas or liquid. The condition manifests itself most noticeable when calculations are performed during the dehydration stages of the wallboard heating. Once the dehydration of the boards is complete and any liquid water has evaporated, the mass transport equations become almost trivial and unconditional stability is recovered. We use a variable time step where, at the completion of a calculation, the results are checked to assess whether the condition expressed in (40) is violated. If it is, the time step is reduced and the solution recalculated. If the condition in (40) is not violated, a new time step is determined and the solution proceeds.

Model Parameters

Of as much importance as the equations are the thermal properties of the gypsum board utilized in the model. The parameters utilized are presented below with a discussion of the source from which quantities were obtained.

(1) Density

The density of the Type-X gypsum board utilized in the simulation was obtained from the measurements of Mehaffey et al [9]. This value was chosen as this was the source for the furnace test that the model presented will be used to simulate. The measurements further showed that the core of the 0.0159 m board lost approximately 17.5 % of its mass during heating. From this value, the density of the core material for each state of the $CaSO_4$ n-hydrite was calculated by assuming that mass loss only occurred through dehydration of the gypsum. The calculations predict that the cores were initially composed of approximately 84 % $CaSO_4 \cdot 2H_2O$. With these values, the density of the gypsum wallboard for each state of calcium sulfate is

State #	Density (kg / m3)
1	648.
2	570.
3	534.
4	534.

(2) Specific Heat

A study is currently underway to measure the specific heat of core material utilized in gypsum wallboard as well as describe the dehydration mechanism [7], but is at present still incomplete. For the development of the model, therefore, specific heat values were obtained from handbook results [11] by assuming that the entire core was composed of initially $CaSO_4 \cdot 2H_2O$.

State #	Specific Heat (J / kg K)
1	1080.
2	803.
3	703.
4	703.

(3) Thermal Conductivity

The thermal conductivity utilized was obtained from data presented in [5], and shown in figure 4. While conclusions may be reached concerning the state of the calcium sulfate for the measured values, incomplete information exists for describing the thermal conductivity across the entire temperature range for all states of the calcium sulfate. We therefore chose to assume the

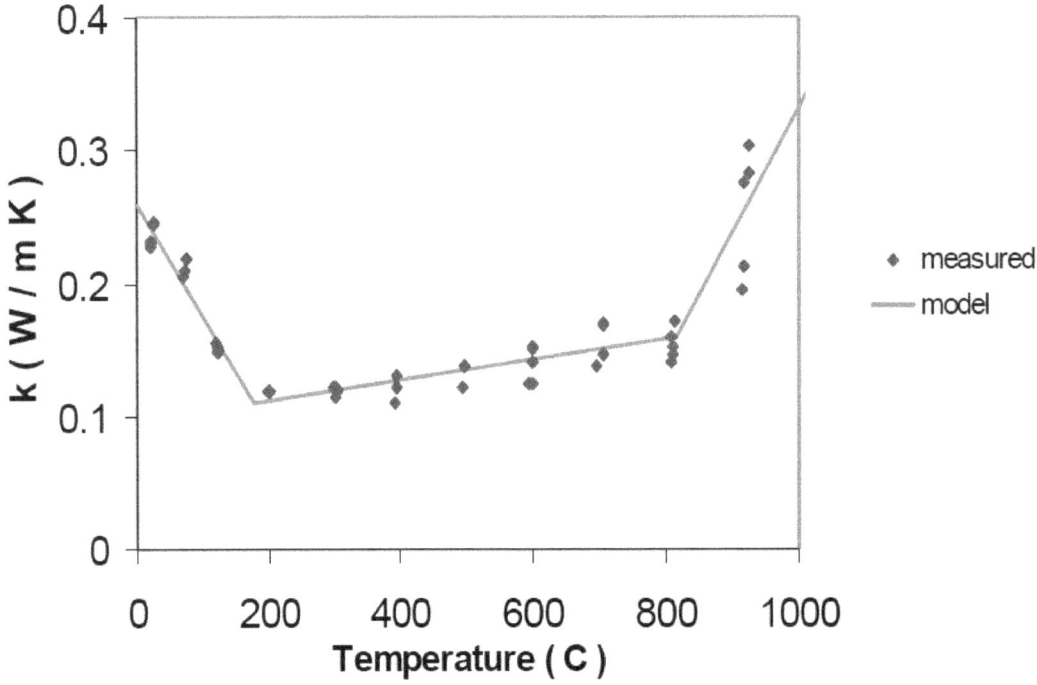

Figure 4: Measured [6] and as modeled thermal conductivity of gypsum wallboard.

temperature dependent thermal conductivity presented as the solid line in figure 6 for all states of the core.

(4) Porosity

The porosity of the gypsum board was estimated from mercury injection measurements presented in [12]. In this study two different gypsum board samples, one obtained from France and one from Denmark, were tested. The similarity of the results provides at least some confidence as to the porosity of the boards utilized in the experiment. The porosity value assumed for the gypsum board was $\varphi_0 = 0.3$.

(5) Diffusivity

The diffusivity of water vapor across a gypsum board has been found to depend upon the amount of free or absorbed moisture present in the board as well as on the temperature of the board [13]. We chose to model the diffusivity by using the 0 % moisture content diffusivity presented for measurements at 6.7 °C (44 °F). The presented governing equations then implement their own model for the dependence of the diffusivity on the amount of free condensed water present in the board by including the pore fraction available for the gas f_g. Temperature and pressure dependence is modeled by assuming ideal gas behavior and estimating the diffusion coefficient from the relation [8],

$$\frac{D}{D_0} = \left(\frac{T}{T_0}\right)^{3/2} \frac{p_0}{p} \tag{41}$$

where the subscript 0 terms represents the reference state. Both the temperature and pressure values used in this expression are absolute. The reference value of the water vapor diffusivity was $D_0 = 9.03\text{E-}09 \text{ m}^2 / \text{s}$.

(6) Permeability

The permeability of gypsum board was calculated from the air flow resistance measurements of Bassett [14] for gypsum plasterboard. From these results the permeability is estimated from the relation,

$$K = \frac{\mu l}{R} \tag{42}$$

where μ is the viscosity, l is the thickness of the board and R is the air flow resistance measured at a given pressure differential. The calculated permeability for gypsum board was taken to be $2.0\text{E-}14 \text{ m}^2$.

(7) Adsorption

The adsorption of water to the pore surfaces was estimated from the isotherms presented in [8]. As discussed previously, no temperature dependence was incorporated on the adsorption mechanism. The functional form of the correlation used to express the isotherm is,

$$A = S \frac{\alpha \phi}{(1 + \beta \phi)(1 - \gamma \phi)} \tag{43}$$

with the utilized parameters: $\alpha = 0.00336$, $\beta = 0$ and $\gamma = 0.9010$.

(8) Reactions

The parameters used in the reaction expressions are estimated from early analytical results that will be presented in a future publication [7]. The values were obtained by decomposing apparent specific heat curves similar to that presented in figure 1. From these values and the Arrhenius model, an estimate of the parameters can be made. For the current model, reactions were assumed to be only of order 1. The values for A and B from equation (16) used in the calculations presented here are,

Reaction #	A	B (K)
1	-3.56E30	31 700
2	-1.93E21	25 100
3	-8.51E23	41 630

Heats of reaction were calculated from handbook enthalpies of formation [11]. While it was possible from the apparent specific heat curves to estimate the heat absorbed or released during the chemical reactions, these values required imposing temperature dependent specific heat values. This level of detail was not desired in the initial formulation of the model. The heats of reaction used in the presented calculations are,

Reaction #	Δh (J / kg of state i)

1	-490 000
2	-209 000
3	78 000

(9) Gas Properties

When constitutive properties are required for the gas mixture, it is assumed that the entire mixture is simply air. Thus, for example, in the implementation of Darcy's Law for the flow of gas through the porous material, the viscosity utilized is that for air. Temperature dependence of the properties was obtained from the appendix tables of [10].

(10) Water Properties

Viscosity, specific heat and thermal conductivity values for liquid water were obtained from the appendix tables of [10]. Enthalpies of evaporation were obtained from the steam tables of [15]. Due to the nearly incompressible nature of liquid water, the density of liquid water was assumed to be independent of pressure but dependent on temperature with the density values taken from the saturated steam tables of [15].

Results

We choose to simulate the results of a fire resistance test originally presented in [4]. The test simulated, test #6, is a full scale test of 0.0159 m thick Type-X gypsum board on either side of 0.089 m wood studs spaced 0.4 m on center. The furnace exposure of the test matches the time / temperature curve of ASTM standard E119. Surface temperature results are presented in figure 5 for the experimental measurements and model predictions along with the ASTM furnace temperature curve.

As can be seen from the figure, the results of the model under-predict the surface temperature as compared to the experiment with a few exceptions. The qualitative response however is promising as we see that many of the features of the experimental measurements are recovered in the model calculations. The model results show that the internal surface of the exposed board heats up and then levels off for a brief amount of time. This portion of the curve corresponds to the presence of condensed liquid water present within the board between the fire and the internal surface. The presence of the liquid water acts as a temperature regulator for the unexposed surface. Once the water fully evaporates from the board, the temperature then begins to rise again. Similar behavior is also observed on the ambient side of the unexposed board, with the additional feature of a slight cooling of the board as the water evaporates. Again, once the water fully evaporates from the board, the temperature again begins to rise. While the results of the simulation have only been presented for 60 minutes, the simulations were actually allowed to proceed for a 75-minute exposure. At the completion of the simulation, conditions had still not yet been reached that would represent an insulation failure for the partition, that is 139 °C over the initial temperature. While the experiment considered above failed structurally at 50 minutes, we can compare the simulation results with a similar test conducted for Type-X gypsum wallboard on steel studs with a time to insulation failure was 52 minutes [16]. This comparison illustrates the models under-prediction of the rise in temperature.

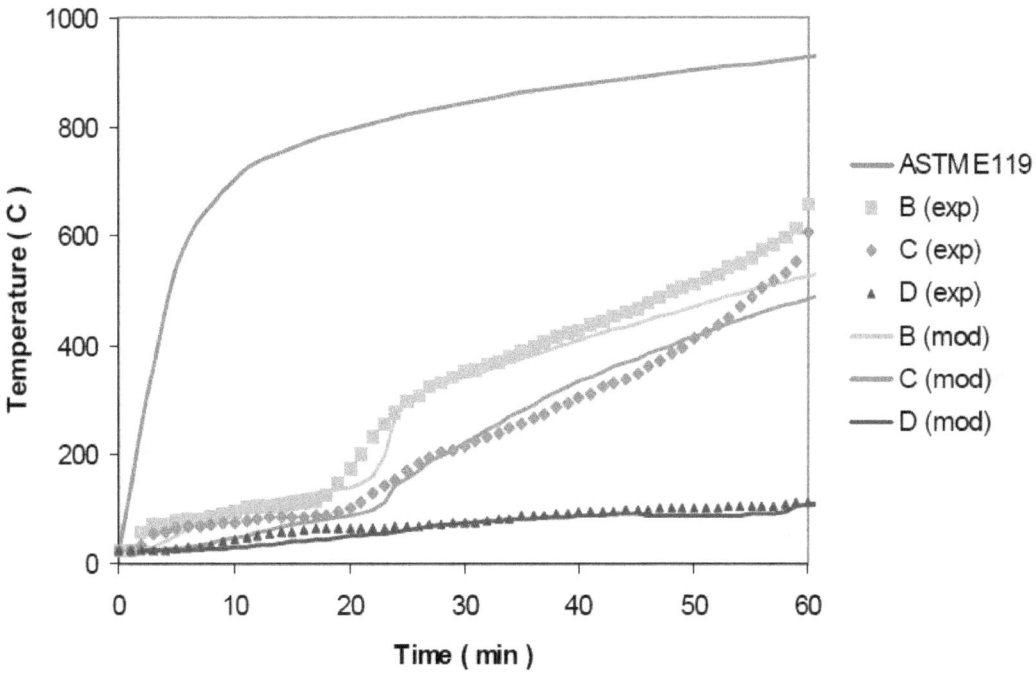

Figure 5: Surface temperature comparison between experimental measurements [4] and model predictions. Points B, C and D correspond to labels in figure 2.

Quantitatively it appears that insufficient heat is being transported across the boards especially during those times when the temperatures are low. During the experiment, the temperature on the interior surface of the exposed wallboard was already approximately 60 °C two minutes into the test. By five minutes, the temperature on the interior surface was approximately 80 °C. This rapid rise cannot be explained through simply conductive effects and therefore points to an additional heat transfer mechanism. It was, in fact, this behavior that drove the initial derivation of the current model, although it appears that the mass transport in its current form is unable to produce the initial heating. It should be noted that the rapid early rise in temperature has been observed in the presentation of results from other similar fire resistance test of gypsum partitions as well, c.f. [2-4,9].

Even though the model does not accurately predict the response of the gypsum board, results of the calculations do provide some insight as to how the partition responds to the applied heat flux. In figure 6, the total amount of condensed (free) liquid water present in the boards in plotted as a function of time. Initially, there is not any liquid water present, however shortly into the test, less than 5 minutes, a condensation layer begins to form in the exposed board. The amount of liquid water then increases as heat continues to be transferred across the board until approximately 11 minutes into the test at which point the amount of water rapidly diminishes. The thermal conductivity of liquid water at 100 °C is 0.68 W / m K and is approximately four times the

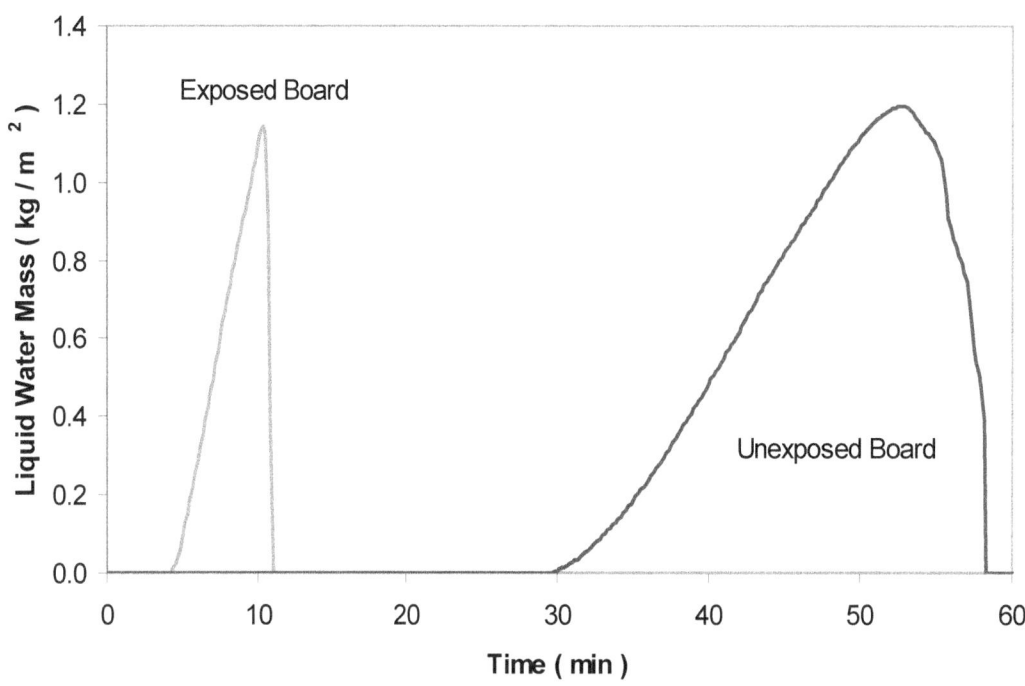

Figure 6: Total liquid (condensed) water mass per unit area in gypsum boards during furnace exposure.

thermal conductivity of the underlying solid material. Therefore, when liquid water exists within the pore spaces, a greater amount of thermal energy can be conducted through these spaces. The rapid heating of the exposed board along with the presence of the liquid water implies that a significant amount of water, more than that currently predicted, is condensing out of the gas into the pore space. A similar, albeit less severe, result is observed in the unexposed board as well. For this board, the slower rate of temperature rise resulted in there being liquid water present over a longer time, that is the condensation front moved at a slower rate, although the maximum total amount of liquid water present was consistent with that of the exposed board.

The condensation front that arose in the model remained thin for a majority of its existence. For the exposed board, the front initially began in a single computational cell at approximately 250 seconds into the simulation. The front then grew only very slowly as it progressed through the board. At the point at which it reached the unexposed face, approximately 700 seconds into the simulation, the thickness was approximately 1 mm, or 6 % of the total board thickness. While the heat transfer through the cells that experienced condensation was significantly greater than those in which there was not any liquid water present, the small thickness of the condensation layer does not allow for sufficient energy transport to match the experimental measurements. This can be observed more clearly if we compare the results of the above simulation with a similar calculation where we neglect mass transport effects. In figure 7, the surface temperatures are

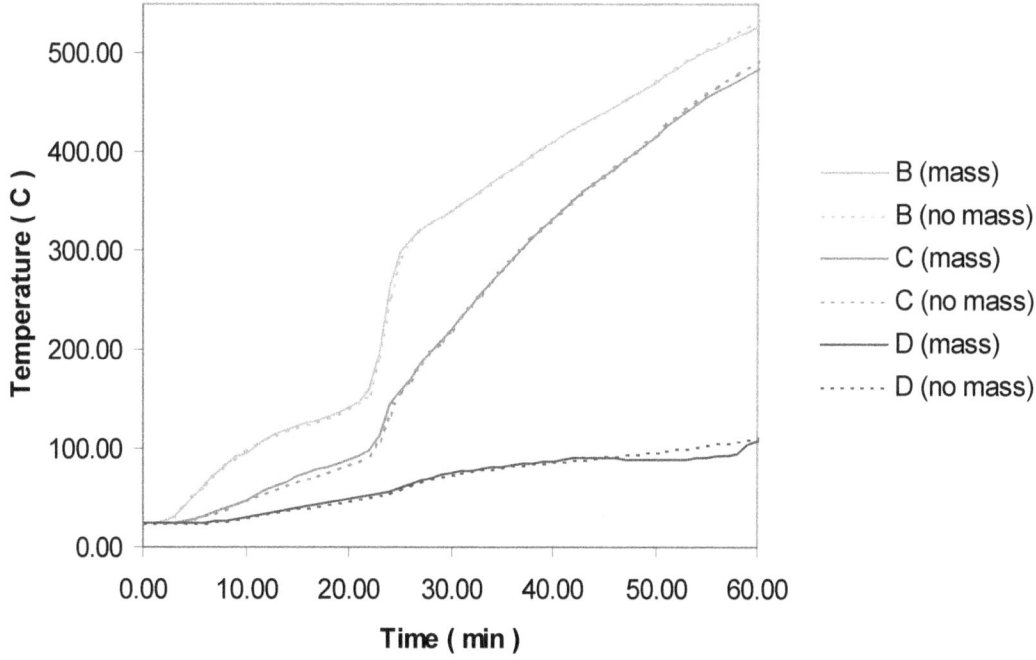

Figure 7: Surface temperature comparison of simulation including mass transport with simulation neglecting mass transport.

plotted versus time for simulations including mass transport effects and simulations neglecting mass transport effects but with all other properties kept equivalent. The figure shows that while there is some differences between the two simulations, most noticeably over the time ranges where the condensation front is at or near the corresponding surface, generally speaking the results are very similar due to the thinness of the condensation front.

In figure 8, the mole fraction of liquid water present in the air space is plotted as a function of time. Initially, a very small portion of the gas contained between the boards is water vapor. As the boards heat up, and the dehydrated water migrates, the gas eventually consists almost entirely of water vapor. This may play a role in describing the transfer of heat across the gas layer. As water vapor absorbs a considerable amount of radiative energy, the assumption that the gas layer between the boards is transparent to radiation should be revisited, especially at later times of the exposure. Furthermore, that the gas layer is composed of almost entirely water vapor means that although the paper backing of the gypsum wallboard will still pyrolyze as the temperature increases, there is a lack of any oxygen to support combustion of the products of the pyrolosis. Clearly, this fact will only remain consistent until a crack or other entry path opens through the wallboard from the inner gas layer to either the fire exposure or to the ambient.

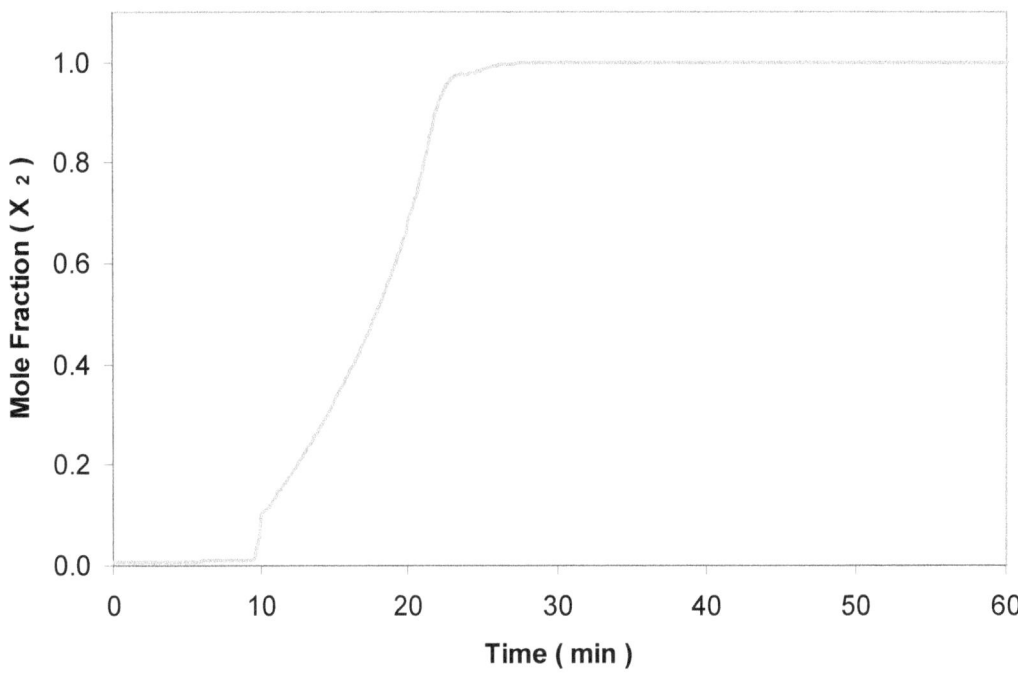

Figure 8: Water vapor mole fraction in gas layer between gypsum wallboards.

Conclusions and Future Work

A model has been presented for describing the heat and mass transfer through gypsum partitions subjected to fire exposures. The model, while still in the development stage, under-predicts the heat transfer across a partition assembly, but exhibits qualitatively similar behavior to that observed during experiments. The model is further able to yield information not only regarding the temperature of the gypsum board, but also information regarding the molecular form of the $CaSO_4$ present in the board core. This information may be useful in determining conditions for crack formation as the board undergoes both heating and cooling during a fire exposure. An analysis of the results yields further information as to the importance of incorporating the effects of the mass transfer brought about by the dehydration of the gypsum boards especially at early stages of the exposure when heat is rapidly transferred across the board as evidenced by the rapid rise in temperature on the inner face of the exposed board.

While the model does not yet accurately predict the thermal response of gypsum boards to fire exposures, the results obtained so far point to improvements and enhancements for the model. First and foremost in this list is the mass transport model. As stated above, the rapid rise in temperature of the internal face of the exposed board implies that an additional heat transfer mechanism, not included in the present model, must be present. While mass transfer has been included from pressure or expansion driven effects, other mechanisms may still exist that drive flows through the gypsum board. The most likely mechanism not included is the potential for capillary flow through the pore spaces of the board. This is currently being looked at for possible inclusion.

Another issue that needs to be considered relates to the transfer of heat across the gas layer. The current model assumes that any gas present within the layer is transparent to radiation emitted from the wallboards. As the wallboards dehydrate, however, the gas layer becomes almost entirely composed of water vapor. The absorption spectrum of water vapor may make the transparent gas approximation no longer valid. This will depend of course of the optical depth of the gas.

As much as was possible, the values chosen for the physical parameters present with the model were selected to represent, from the literature, those most likely to describe the material. Small changes in these values could, however, affect the thermal response of the wall construction, especially under long duration exposures. A sensitivity study should be conducted, with the current model, to assess the relative importance of the physical parameters. This may provide additional guidance regarding the capabilities and the gaps present with the model.

Lastly, the current model treats the wallboard's structure as remaining fixed throughout the exposure and predicts only the thermal response of gypsum. Other effects must also be incorporated if the model is to describe all of the possible failure mechanisms that may arise when gypsum partitions are exposed to fire. Some effects that could be included are cracking, crazing, shrinkage and ablation of the wallboard material.

References:

[1] American Society of Testing and Materials, *Standard Test Methods for Fire Tests of Building Construction and Materials*, ASTM Standard E119-00a, West Conshohocken, PA (2003).

[2] Thomas G, "Thermal Properties of Gypsum Plasterboard at High Temperatures," *Fire and Materials* **26**, pp. 37-45 (2002).

[3] Collier P C R and Buchanan A H, "Fire Resistance of Lightweight Timber Framed Walls," *Fire Technology* **38**, pp. 125-145 (2002)

[4] Takeda H and Mehaffey J R, "Wall2D: a Model for Predicting Heat Transfer through Wood-Stud Walls Exposed to Fire," *Fire and Materials* **22**, pp. 133-140 (1998).

[5] Ramachandran V S, Paroli R M, Beaudoin J J and Delgado A H, *Handbook of Thermal Analysis of Construction Materials*. Noyes Publications, Norwich, NY (2003).

[6] Benichou N, Sultan M A, MacCallum C and Hum J, "Thermal Properties of Wood, Gypsum and Insulation at Elevated Temperatures," *NRC-CNRC IR-710* (2001).

[7] Kukuck S R and Prasad K, "Specific Heat and Dehydration Reactions of Gypsum Wallboard," (in preparation).

[8] Richards R F, Burch D M and Thomas W C, "Water Vapor Sorption Measurements of Common Building Materials," ASHRAE Transactions **98**, part 2 (1992).

[9] Mehaffey J R, Cuerrier P and Carisse G, "A Model for Predicting Heat Transfer through Gypsum-Board / Wood-Stud Walls Exposed to Fire," Fire and Materials **18**, pp. 297-305 (1994).

[10] Incropera F P and De Witt D P, *Fundaments of Heat and Mass Transfer*. 2nd edition, John Wiley & Sons, New York, NY (1985).

[11] Lide D R, *Handbook of Chemistry and Physics*. 71st edition, CRC Press, Inc., Boca Raton, FL (1990).

[12] Blondeau P, Tiffonnet A L, Damian A, Amiri O and Molina J L, "Assessment of contaminant diffusivities in building materials from porosimetry tests," Indoor Air **13**, pp. 302-310 (2003).

[13] Fanney A H, Thomas W C, Burch D M and Mathena L R, "Measurements of Moisture Diffusion in Building Materials," ASHRAE Transactions **97**, part 2 (1991).

[14] Bassett M, "Air Flow Resistances in Timber Frame Walls," *Proceedings of the New Zealand Workshop on Airborne Moisture Transfer*, Wellington, New Zealand (1987).

[15] Sonntag R E and Van Wylen G J, *Introduction to Thermodynamics, Classical and Statistical*. 3rd edition, John Wiley & Sons, New York, NY (1991).

[16] Sultan M A and Lougheed G D, "Results of Fire Resistance Tests on Full-Scale Gypsum Board Wall Assemblies," *NRC-CNRC IR-833* (2002).